GROCERY ROW GARDENING

Books by David The Good

General Gardening
Start a Home-Based Plant Nursery

Florida Gardening
Totally Crazy Easy Florida Gardening
Create Your Own Florida Food Forest
Florida Survival Gardening

The Good Guides
Compost Everything
Grow or Die
Push the Zone
Free Plants for Everyone

Jack Broccoli Novels
Turned Earth
Garden Heat

GROCERY ROW GARDENING

THE EXCITING NEW PERMACULTURE GARDENING SYSTEM

BY DAVID THE GOOD

Grocery Row Gardening
David The Good

Copyright © 2021 by David The Good.

All rights reserved. No part of this publication may be reproduced, distributed, or transmitted in any form without the prior permission of the publisher, except as provided by copyright law.

Good Books Publishing
goodbookspub.com

ISBN: 978-1-955289-07-8

Contents

Introduction		i
1	How To Build a Grocery Row Garden	1
2	Why Grocery Row Gardens Work	25
3	Maintaining a Grocery Row Garden	35
4	The Influences	47
5	Concluding Thoughts	57

To Steve Solomon, my Christian brother and gardening mentor. Your relentless mind and uncontainable spirit are an inspiration to us all.

Introduction

The idea behind this little book is so awesome I'm almost afraid to write it down. The system works so well it feels like it's too good to be true. It's too beautiful, too productive and way too exciting. There must be a catch! Just in case I find it, I was going to wait a few more years before publishing anything other than articles and videos, but, well, I can't help it. I had to write it down and enlist you to join me in this marvelous new permaculture gardening system. My enthusiasm is bubbling over and needed to be printed in black and white so I can share what I've discovered.

Imagine combining the order and efficiency of row gardening with the diversity and beauty of a food forest. That idea came together in my mind after years of experimentation and soaking in the knowledge of a wide range of gardeners.

From 2010–2016, I built and maintained two different food forests in two different parts of Florida, playing with plant guilds and species and bringing order out of chaos. I also assisted with the design and installation of many other food forests across the state. We figured out what worked and what didn't, building soil and reaping sweet harvests of fruits, vegetables, roots and herbs.

In 2016 our family moved to the island of Grenada in the West Indies. Its volcanic soil, tropical but mild climate and good rainfall made it perfect for experimenting with tropical crops. Unfortunately, it took a long while to pin down and purchase a piece of land. In the interim, we rented and tested crops on borrowed land. When we finally got our own land, we spent our first year of ownership in building a place to stay. During that time, we cleared most of the gigantic overstory trees and planted a traditional food forest on the top 1/3 acre of the land. We also hired an excavator to re-dig an old drainage ditch across the bottom 1/3 of our land after neighbors warned us that it was prone to flooding.

This bottom 1/3 of the land was rich volcanic clay loam. It was also full of rocks and some large boulders. Over a month or so we managed to remove most of the giant boulders from the ground by carefully prying them up with a Meadow Creature broad fork, then rolling them to the edges of the land.

With the large trees gone and the boulders moved, we had space for a traditional garden. But I wasn't quite happy with a traditional garden anymore. I wanted something better. We started by planting pak choi and tomatoes, cassava and yams—but there were greater things coming.

For years I had planted food forests and annual gardens in separate areas, with some exceptions. Back on our North Florida homestead my daughter Daisy and I planted a perennial garden bed in the middle of our annual gardens as a sanctuary for predators and pollinators. That was in the winter of 2014.

That bed contained three roses, three varieties of raspberry, rosemary, garlic chives, lion's ear, perennial marigold, oxalis, cut-leaf coneflower, milkweed, plus a Saijo astringent Japanese persimmon right in the middle.

It was gorgeous and grew fine in the midst of my other gardens. Yet there were a few considerations that held me back from planting trees and shrubs all through my annual garden beds. First of all, I was concerned about the shade cast by trees reducing my yields. I also did not want to give up the ability to till and re-shape beds. Over the course of a decade, the space had already transformed from a 25' x 25' anarchistic block of crops with vein-like pathways winding through, to a set of uniform wood and cinder-block beds, to an open patch of mounded earth beds with some cinderblock beds around the edges. I kept playing with different designs and crops, particularly as I performed the crop experiments that eventually led to writing my book *Totally Crazy Easy Florida Gardening*.

One day I might be setting up an old bed frame to grow beans over. Another day I might be transplanting papayas for a zone-pushing experiment. I had already locked up my entire front yard with a half-acre food forest. I needed that sun in the back gardens to keep my annuals growing. How else would I get enough light to breed Seminole pumpkins, plant Jicama and try experiments with cassava and sweet potato intercropping?

Yet still, trees started to creep in, especially as I realized the power of pruning to keep them under control. When I finally got it through my head that the "maximum height" of a tree was not an inevitable outcome of planting said tree, the lights came on. The trees did not need to overshadow all my annuals. They didn't need to consume all the resources in an area. They didn't need to become a canopy. With pruning I could put them anywhere, even in the middle of an annual bed.

With great enthusiasm, I researched all kinds of tree-growing literature, from espaliering to pollarding. As I experimented and observed, new lights kept going on in my head. I read all I could on edible hedges, I devoured Julia Morton's tropical fruit tree profiles, I watched Stefan

INTRODUCTION v

Sobkowiak's film *The Permaculture Orchard*—and I kept thinking and experimenting.

When I moved to the tropics, I learned and observed and planted and tested across our multiple rental properties before finally getting our land—and I was almost ready to create the system that was building inside my head. It would be a combination of the best of annual gardening with the best of food forests with the best of an orchard. A synthesis of multiple ideas, solving multiple problems and building upon all that I had learned.

Once I had my land, I was closer. And when my annual garden area was cleared, I was itching to try something new. In a livestream I discussed the idea for making garden beds that combined trees with perennial and annual crops. In response, Andrew Cavanagh, who had seen my stream, sent me a ton of information on a tropical gardening system called "Syntropic Farming", or "Syntropic Agriculture". This system had been developed over multiple decades by a Swiss farmer living in Brazil by the name of Ernst Götsch. It's dense and heady stuff. As his website relates:

> *[Syntropic Agriculture's] creator, Ernst Götsch, bases his worldview on a transdisciplinary scientific approach and a practical daily routine for more than five decades. The logic that guides his decision-making process follows a*

path that arises from Kant's ethics and crosses physics, Greek philosophy, and mathematics. It also relies on biology, chemistry, ecology, and botany, and incorporates the current technological scene, adapting techniques and tools from other areas. Ernst Götsch's agriculture relies on a coherent and systematic chain of data, free of internal contradictions, which is not only sustained by a logical narrative but also includes a practical and concrete expression at the end. From planning to planting, there is a method, and there is a practical result. More than a good idea, Syntropic Farming has proven to work, and it can address the biggest social and environmental challenges of our time. (agendagotsch.com/en/what-is-syntropic-farming/)

Whew. Practically speaking, Syntropic Farming plants perennials and annuals together in rows and uses chainsaws liberally to keep the canopy from closing and to feed the crops. There are cycles of planting and succession that add new crops at specific timing to ensure the system remains continuously productive. It's a design marvel and is an amazing way to use tropical agriculture for high yields without destroying the soil. Bananas and plantains feature prominently in the system, providing both a crop and a regular source of biomass to feed the soil.

INTRODUCTION

After reading through the literature in English, I realized this system paralleled much of what was brewing in my mind. I built a bed and incorporated some of Ernst Götsch's system in my gardens and really liked how he used bananas and other biomass producers to feed short-term crops. His work straddles the line between annual bed-based agriculture and food forest systems—a line I was already bending. These ideas were dumped into the Good Gardening hopper with my own research and experimentation alongside the knowledge I'd gained from many others over the years.

The Syntropic system, though incredible, takes way too much organization for the average gardener. My gardening motto is "more food for less work", and I am allergic to complicated systems that require precise timing and specifications. As I looked at my lovely new gardening space in the jungle, I decided to blaze my own trail.

We launched the first iteration of Grocery Row Gardening in my Grenada backyard at the beginning of 2020, creating mounded well-dug garden beds of 4' width with 2' wide paths in between, planting them with trees, shrubs, herbs, vines and vegetables in a system of perennials and annuals spaced wide enough to live without much irrigation.

And then Coronachan came knocking, locking down our island and shutting almost all the stores. Groceries were almost completely closed and there were long lines for food.

People were panicking and we were isolated on our little half-acre homestead in a foreign country and basically forced to "Grow or Die". Before the lockdown I had seen early reports of the virus in Wuhan and had purchased a bunch of storable food along with a good ration of rum—yet the lockdowns went on and on and shopping was very difficult, so our supplies dwindled as the spring melted into summer. It was a great test of our new gardens. Much to my pleasure, the system worked well, providing us with an abundance of vegetables, beans, roots, tomatoes, peppers, okra and herbs. The bananas and young trees grew quickly in the rich and well-tended beds, and it looked like 2020 was going to be a banner gardening year. I could almost taste the bananas and the papayas that would be coming in later in the year.

But the attitudes on the island had shifted. Our neighbors no longer liked having Americans on their road. The government would not renew our visas based on my work as they had in previous years. Our next-door neighbor would mutter curses as he eyed us across the property line, no matter how kindly we acted. We had done our best to fit in, sharing seeds and even paying to dig out his ditch when we had ours done. It didn't help. When things got stressful during the virus, we were unwelcome and it showed. Beyond that, some locals were talking openly of sticking up and robbing "rich people" as the lockdowns continued and we knew that meant us, no matter how simply we lived. (At this point we lived in a pair of tiny houses with an outdoor shower and a composting toilet. We were not rich, but we were Americans and that was enough.)

So we left. We put our land up for sale, said goodbye to our friends and gardens, and flew back to the States. It had been a good run and we learned a lifetime's worth of information about living cross-culturally and gardening in the tropics. And I had hatched the beginnings of a great gardening system.

We hunted for a rental in my home state of Florida without luck, then ended up finding a place in Lower Alabama by God's intervention. We never expected to be here and had never even visited Lower Alabama before, but a friend of a friend had a house with some land and offered it to us at an affordable rental price at just the right time. God is good. Now we had a place to garden again as well as a place to rest our heads—the former being no less important than the latter, of course.

I was so psyched up about the Grocery Row Gardening concept, I welcomed being thrown into a new climate.

Unlike the fertile soil of Grenada, the dirt at our new place was so poor it could hardly be called "soil". The surrounding woods were a mixture of pines, popcorn trees, various blueberries and their relatives, scrub oaks and lots of yaupon holly. The grass was thin and patchy and even the weeds weren't that happy. Instead of volcanic clay-loam, we had acid sandy grit with gravel in it. It drained fast but was compacted and airless a few inches down.

I laughed at the good work of Providence. This was perfect! It was one thing to have great success with a garden when the weather was always in the 70s and 80s and the soil was incredible. It would be another thing to have success in a place that suffered deep freezes in winter and temperatures into the 100s during the summer, with soil so bad that it would barely support weeds.

If the Grocery Row Garden system worked here, it would work anywhere. Plus, I got to plant temperate trees again, like peaches, apples and plums.

As soon as we began renting, I started tearing up the sand and planting new gardens. Now, most of the way through 2021, I can tell you that the system works here too, even with lousy soil. We now have a new garden even more beautiful than the the one we built in Grenada, despite its sad lack of banana trees. It's also super productive. I tweaked the original design based on what I learned in 2020 and changed the species I grow to fit a temperate climate—and it's all growing and producing, even in the brutal heat of summer.

Grocery Row Gardening evolved out of a combination of deep research and laziness. It takes the best of many complicated systems and makes them usable for normal people working in normal backyards with normal schedules. But—that said—I don't feel completely confident writing a full book on the system yet. It's only been two years in the testing. Yes, the Grocery Row Gardens are based on huge amounts of previous experimentation, crystallized into a synthesis that looks very solid. Yes, there's no reason that I can see why they won't work as well as my previous food forest systems, annual gardens, sweet potato beds and plant guilds. It should all work together just fine. Still, I would love to have producing apple trees and buckets of figs before I say "DO THIS!" It would be good to see how the tree canopy fits in over time, and see if I need to tweak my spacings tighter or looser or change some of my planting plans. Perhaps some things won't thrive long-term in the system. Maybe I can't grow pomegranates and blueberries together in happiness due to pH or some other reason. Maybe my Everglades tomatoes are going to overrun the whole system. Maybe a horde of poisonous death toads are going to emerge from the ground and eat my strawberry plants. Maybe, maybe, maybe...

...but I think you can handle it. I believe the joy of experimentation in your heart is much stronger than your fear of failure. I was asked multiple times to write a book on this crazy beautiful mad lovely gardening system that people have seen me build on my YouTube channel, so I'm writing it. Consider this my "white paper" on the Grocery Row Gardening concept. It's a snapshot of my thinking right now, along with what I imagine for the future and what I've already discovered about what works and what doesn't. I know this system works great for annuals. I know that the perennials and trees we've planted so far are growing excellently. I also know I can keep fruit trees small and productive.

I firmly believe this will work even though I don't have a decade-long track record like I do with traditional food forest design. I am sure it is both an improvement on the traditional food forest and the annual garden. I know it solves many problems faced by gardeners as well as creating one of the most beautiful gardens you can imagine.

If you can overlook my shortcomings and the youth of this line of experimentation, I invite you to come along with me and enjoy the wonder of creating your own Grocery Row Garden. In a couple of years I hope to expand this little book as I harvest pears, mulberries, persimmons and grapes—but for now, let this be my initial inscription on the value of this new method of gardening.

Onwards and upwards!

David The Good
Atmore, Alabama, 2021

Chapter 1

How To Build a Grocery Row Garden

A Grocery Row Garden is like a food forest and a kitchen garden arranged in strips with paths in between. It's a form of agroforestry but it's also an annual garden. In it, you'll be growing trees, shrubs, perennials, annuals and all kinds of wonderful things in a profusion of species and lush growth that is naturally bounded by controllable paths. Grocery Row Gardens unleash the crazy abundance of a food forest inside the comfortable control of a simple annual garden bed.

Step 1: Make the Beds

The Grocery Row Garden system starts with one or more mounded beds of soil. The beds are 4' wide and as long as you like, with 3' wide paths in between.

I know. 3' wide paths seems excessive. But there is a reason behind the madness.

If you have a very small yard, you might be tempted to make your paths really tiny—maybe even 1' wide—but I don't recommend doing that. I've made the mistake and it makes for an awkward garden. The narrowest I would go is 2' in width, especially with Grocery Rows. After seeing how crazy the plants have grown this year, I might even consider using 4' wide paths between rows. Whatever you do, don't skimp on the path size.

When I first watched Stefan Sobkowiak's film *The Permaculture Orchard*, the amusing idea of "backyard shopping" really entered my mind. The idea of pushing a grocery cart through a garden that stretched upwards on either side of the "shopper" struck me as a fun way to think about the garden. Could you pick things from different layers in a garden? High up, in the middle, and down low? Would you get a great mix of produce?

I call it Grocery Row Gardening because this garden is rather like a supermarket, where rows of shelves can be accessed by shoppers moving between them with carts. You go down one row and pick up some canned beans, some apple juice, a few bags of beans and a bag of rice, then move on to the next row, where you get some sardines, pickled onions, sauerkraut, and breath mints, etc. I first titled the system the much more complicated name of **NOGRASS** gardening, AKA the *Natural Organic Grocery Row Agroforestry Soilbuilding System*, but abandoned that name because I do use some grasses as cover crops. Plus, the system isn't truly "organic" by current standards since I use some elemental fertilizers. *(And that name is a pain to remember, so forget I ever called it that, okay? It never happened. I don't even know why I'm bringing it up. It's definitely always been called Grocery Row Gardening.)*

If there is a genesis to the "Grocery Row Gardening" name, it's probably Stefan's permaculture orchard in Quebec. I've been picking groceries from the backyard since I was a kid, but Stefan really went "up" with the idea and was a great inspiration. This system isn't a commercial system like his orchard, however, and has more emphasis on annuals, as well as shorter trees and tighter rows, but I definitely owe him a debt of gratitude for the "shopping" idea.

Each aisle in a grocery store has a wide selection of items for sale. A Grocery Row Garden is similar except you can take what you like without paying a cashier. You have fruits, nuts, berries, vegetables, roots, herbs and flowers all along the rows, ready to be picked as they ripen. If you limit the space between rows too much, caring for the plants and harvesting the rows becomes a pain. You should be able to walk past other people in a row without bumping into them.

It's also good to be able to bring through a wheelbarrow or a cart so you can mulch or take out a few baskets of tomatoes without running over lots of vines along the way. Happy plants in these beds will spill onto the paths and try to fill them over the course of a season. We've already had this happen with tomatoes and strawberries, summer squash and the loathsome vegetable that starts with the letter "z", sweet potatoes and tomatillos. The branches of trees and the canes of berries also spread out into the paths, so if the paths are already narrow, they become well-nigh impassible by summer. You don't want to harvest Big Boy tomatoes while pinned between the Scylla of raspberry thorns and the Charybdis of wandering squash vines. My first Grocery Row Garden in Grenada had 2' wide paths, which was fine for beds of pak choi and tomatoes but was a little inadequate when bananas, cassava and other aggressive plants were added to the mix. When I moved to Alabama, I added a foot to all the paths when I planned the new system, and so they remain.

As for the 4' wide beds, I have found that to be the optimal balance between growing space and accessibility. You can jump a 4' bed and work it easily from both sides with the reach of an average human arm. The same is not true of a 5' bed. Sorry, John Jeavons—I've never been able to get behind the 5' bed. You get more growing space but you sacrifice comfort and accessibility. Sure, a Grocery Row Garden with

2' paths and 5' beds would give you a lot more growing space but it wouldn't be nearly as enjoyable to grow in. It's also not as good for the trees. The skeleton of a Grocery Row Garden is a well-spaced orchard. You're just superimposing a beautiful annual garden and a bunch of perennials on top of it, letting your vegetables be an understory to the trees instead of grass as in a traditional orchard. It gets more complicated than that, but it's really a matter of planting trees at good spacing and then working around those arboreal tent poles to erect the rest of the structure.

In order to make laying out Grocery Row Gardens an easier process, my 8-year-old son and I made some measuring sticks from 1"x1.5" scrap wood, cutting them at 3' and 4' lengths and painting them different colors, while also writing the length on each stick with a marker. The 4' one is for the bed width, the 3' for the path width. We also made a 2' for measuring crop spacing inside the beds but it isn't used in the creation of Grocery Rows.

Make sure you choose an area in full sun for your new garden. Most of your plants will need the sun, especially since there will be some extra shade as your fruit trees get established. It's also good to stay away from existing trees if you can, as their roots are greedy.

There have been some questions about whether it is better to lay these beds out North to South or East to West. I researched this question when I was laying out grape trellises one year. The best choice depends on your land and your location. As the Cornell Cooperative Extension Lake Erie Regional Grape Program notes:

> *In rows oriented north/south, vines tend to have more even sunlight distribution in our region. Several studies (and plenty of anecdotal evidence) demonstrate east/west rows tend to have uneven ripening or quality issues in regions with cool climates, especially the Northeastern US. Frequently, the site itself will often dictate in which direction rows can be aligned. For example, if a site allows for either numerous, short north/south-oriented rows or longer east/west rows, one should opt for the latter. Above all, however, one must consider the practicality of row orientation and length within the confines of a particular location, especially while considering the necessity of headlands and turning space for vineyard tractors, harvesters, and other equipment.*

Though vineyards aren't the same as Grocery Rows, the ripening of fruit and the sunlight falling on vegetables should be considered, particularly in colder climates. Your layout should also consider the land you're using. If there is a slope, be sure to build your beds perpendicular to the slope and not up and down with it. Paths rapidly become ruts if they run in the same direction as the flow of water during a rain. If the slope is more than a few degrees, you may also want to dig a drainage channel at the top of the gardens to divert some of the water coming downhill. Trees do a great job holding together the soil but a young system can be greatly damaged by a flood, plus running water will wash away a lot of your soil fertility and organic matter. Plan for water right from the beginning.

In a more Southern climate the layout of your rows probably doesn't matter much. You'll get plenty of sunshine no matter which way you run the beds.

Before measuring or building any beds you need to clear the ground you're going to plant. Mow it down, then dig or till the entire area. In Grenada we built the system a bed or two at a time, laboriously slashing down all the weeds and vines and small trees with machetes, removing and burning stumps, prying up boulders and then using a Meadow Creature broadfork and a variety of eye hoes to loosen all the ground and help us remove the roots and rocks. We reclaimed the jungle in pieces, getting a lot of exercise in the process. Once we had our clear ground, we constructed our beds and paths. In my new Alabama gardens, I had the great luck to be able to borrow a tractor with a tiller attachment which made short work of the weeds. I tilled a

Lauren Paolini lays out her new Grocery Row Gardens.

large area, planted it with clover, rye, mustard and daikon, let it grow for a few months, then tilled it all under again and started building beds. That cover crop put some life in the ground before we started planting in earnest. Terrible soil needs all the help it can get!

To make a bed, grab some stakes (we use short pieces of rebar) and hammer in two of them at one end of the new bed, 4' apart. Now pull twine down to where you want to put in the other side of the bed and do the same thing. The strings should be exactly parallel at 4' in width for the entire length of the bed. If the entire area is well-tilled, it should then be easy to shovel out loose soil from the space on either side of the bed—which will soon be pathways—into the bed, giving you a nice, rounded mound of loose soil, 4' wide by however long you choose. Most of my beds are about 60' in length right now. If you have a smaller yard, however, it's no big deal to make 8' beds, or 10' or 20'.

There's no law saying that Grocery Rows have to be perfectly straight. If you want to be artistic, you can run them in circles or swirls or waves or curves—just keep that 4' spacing in the beds with 3' paths, so you don't get too lost in the crazy. I like anarchy in my plantings, but I also like straight lines, so I prefer to stick to parallel rows. They're easier to dig and maintain.

Once you've made your first bed, mark off your second 4' wide bed 3' away and repeat the process until you've made all the beds you're ready to make.

Starting with at least three beds makes sense, as it's a nice, balanced-looking garden with lots of good "edge" space for maximum yields.

Step 2: Improve the Soil

Now that you have your beds in place, this is the time to do some soil improvement. In our gardens we have been experimenting with charged biochar; that is, charcoal we make from branches and then soak in a nutrient solution to "charge" it with the good stuff before tilling it into the ground. So far the results have been encouraging in our poor soil and most of the Grocery Row Gardens have a good bit of char buried in them. I also forked under alfalfa pellets into some of the beds before planting. Alfalfa adds nitrogen and some humus and has a magical ability to boost plant growth. We also mixed up multiple batches of Steve Solomon's micronutrient mix—called "Solomon's Gold"—applying about a quart of it to every 10' of bed. This mix was optimized for my garden, based on a soil test I obtained from Logan Labs. I ordered the "Standard With Extras" test to see what nutrients my soil lacked and found it lacked most everything.

The recipe Steve Solomon gave me to replace the minerals in my poor soil was (per batch):

- 4 quarts cottonseed meal
- 1 quart garden lime (better to use 2/3 quart garden lime and 1/3 quart dolomite)
- 2 cups pelletized gypsum
- 3.5 cups bonemeal
- 2/3 cup potassium sulfate
- 1.5 tbsp borax
- 2 tbsp manganese sulfate

- 2 tbsp zinc sulfate
- 2 tsp copper sulfate
- 1 quart kelp meal
- 1/8 tsp sodium molybdate

Remember that this was perfect for my soil but may not be optimal for yours. A better mix for your own soil can be calculated based on the data in Steve Solomon and Erica Reinheimer's book *The Intelligent Gardener*. It's not easy, though.

To make things even more crazy, I also added these additions to Steve's mix for extra minerals:

- 1 cup Azomite
- 1 cup Sea-90
- 1 cup greensand
- 1 cup magnesium sulfate

If you can't figure out how it works and want nutrient-dense food, I recommend just throwing on some extra kelp meal with whatever your use as fertilizer so you can get all the minerals of the ocean in your produce.

Along with this mix, I added a half-bucket of clumping kitty litter to each 4 x 60' bed. That kitty litter contains bentonite, which will help make my soil "stickier", causing it to hold more water and making the minerals in the ground more accessible to plant roots. I will probably add another bucket or two to each bed in the future as well. If you have clay soil, don't do this! You definitely do NOT need more "stick". Instead, add extra gypsum to loosen the soil.

If you have some compost to add at the beginning, add it. We dug up an old manure pile and some rotten logs to add organic matter to the beds. But be careful! I've warned people many times about the danger of long-term herbicides like Aminopyralid killing gardens that are fed with manure or hay. Don't roast the gardens with poison! Many hay fields are now sprayed with long-term herbicides that kill almost everything except grass, contaminating the hay and whatever manure is produced by the animals that eat it. This stuff sticks around and will kill your trees and plants. Don't take any risks. Someone just wrote me this morning and sent a bunch of photos of their destroyed gardens, pleading for help. They had applied horse manure which had come from a neighbor's farm, killing everything. It's not the manure that's bad—it's the nasty chemical herbicides. Avoid them! The only reason I used the old manure I did was because I knew the farmer was too cheap to buy hay or to spray.

Beyond adding Solomon's Gold and manure, good sources of soil fertility for your beds might include:

HOW TO BUILD A GROCERY ROW GARDEN 13

- Seaweed
- Leaf mould sifted from the woods
- Rabbit manure (if not fed on hay)
- Compost or dirt from a chicken run
- Homemade compost
- Ashes and charcoal
- Alfalfa pellets
- Coffee grounds
- Hair from a barber shop
- Mulch from a tree company
- Grass clippings

Or even just bagged fertilizer. Old Alabama Gardener on YouTube uses triple-13 on everything and does fine. The better your soil is at the beginning, the less it will need to be fed. You can go organic with a Grocery Row Garden or use something like Peter's 20-20-20 as a foliar feed—I won't judge you. The point is to grow food, and I really don't care much anymore what fertilizers people use. The plants don't seem to mind much either way, so long as there is enough organic matter in the ground to buffer the effect of chemical fertilizers, and provided you are providing the plants with good micronutrients. Over time, your beds will get richer anyhow—which I'll explain in a little bit. You just need to get them established at the beginning with whatever you have available that will feed the plants. If you have some mulch, put it down. If you don't, don't worry about it. Hopefully you'll have the ground covered in green vegetables soon. But we're getting ahead of ourselves.

Step 3: Plant Your Trees

This is where things get serious. Trees are a commitment. Are you ready for a commitment? Of course you are! Nothing planted, nothing gained.

Some trees lend themselves to growing in smaller spaces. In Grenada we grew papaya and bananas (which are both fake trees that we pretend are real trees) and coffee inside beds with no problem. They were pretty easy to keep under control. Bananas need to have their clumps thinned and replanted regularly, though, or they'll make a big colony that will eat your paths. Syntropic Farming relies heavily on bananas and plantains for improving the soil with lots of biomass, so they are a good addition—you just need to manage them.

In cooler climates bananas freeze before producing fruit, so if you're past about zone 9 your chances of using bananas as a highly productive part of your Grocery Row Gardens are roughly zero. You can grow the plants, but they'll freeze to the ground in winter and spend most of the growing season re-growing, only to be frozen again in the following winter. There is the possibility of growing *Musa basjoo*, the cold-hardy Japanese fiber banana, as part of the system to feed the ground, Syntropic-style, but it does not make edible fruit. I purchased a half-dozen of them to experiment with in future years despite their non-edible nature. When I install more Grocery Row Gardens I will plant some of them and see how they do. If they work out, I will put more information into the next edition of this book.

If I were still in the Caribbean, I would experiment with other tropical trees in my Grocery Row Gardens, such as acerola, jabuticaba, Surinam cherry and even mangoes. The latter need regular pruning but can indeed be kept small if you stay on top of them.

In my Zone 8 garden I have planted a good variety of temperate-climate trees, including figs, pomegranates, peaches, plums, apples, crabapples, pears, hazelnuts, apricots, serviceberry, and Japanese persimmons. This fall I plan to create more beds containing mulberries and Chinese chestnuts, which somehow got overlooked in the first planting. With a system like this I would avoid planting very vigorous trees that want to get huge, like tamarind, American chestnut, pecans, jackfruit, black cherry and red mulberry, as they're likely to fight you too much to stay well-controlled in such a small space. Basically, the top layer of trees in this system is composed of what would be the second-tier trees in a food forest system. We don't want to let anything get taller than maybe 8' in the Grocery Rows or it will consume too much space and create too much shade. You may experiment with it, of course, as this is all wide open.

The width of the trees inside the bed can vary. You might want to train them a long ways sideways or you might not. It's up to you. In the Deep South the extra shade on the plants below may be welcome. Up north, it might not be. I've seen fan-trained fruit trees that look amazing and would fit well inside a row. Think of an apple tree that looks like a sea-fan and you've got the idea. As for my trees, I plan to keep them more in tighter vase shapes.

Though my trees are being kept small via pruning, I still planted them at a generous 12' spacing so you keep plenty of space open for more than just trees. They are the canopy, but they must not become a dense row of trees or you'll just have a fruit hedge. The 12' spacing also means that if we ever stopped growing annuals in the rows and let the beds revert to grass we'd still have decent orchard spacing.

There is a possibility that the tree roots may become competitive with other plants over time. I've grown a lot under trees in my food forest systems without trouble but

I haven't tested every species. It could be that future root pruning as well as top pruning may be necessary. If so, some chops with a shovel in a line across the bed about 4' away from the trunk of a tree in either direction ought to suffice. Chop down about 12 inches and you should be good for the season. I'd probably do this in early spring.

Once you have your trees planted at 12' intervals in your first bed, it's time to plant the second bed. In that second bed, plant the trees at an alternate diagonal spacing to the first bed, giving them more breathing room than being spaced 7' across from each other on opposite sides of the path. It's like a zig-zag when viewed from above. Alternately, you can simply skip the trees in alternate beds and plant shrubs or vines instead, like rabbiteye blueberries or grapes on trellises. I put shorter shrubs in my in-between beds, alternating a bed planted with fruit trees with a bed planted with shrubs. That way if the pruning is neglected in the future we'll still have a well-spaced orchard. The first farm we rented in Grenada had its entire front yard planted with fruit trees at very close spacing. They had all grown together and become much less productive. The shade also caused the grass beneath them to die, leaving slick mud that was a hazard on rainy days. It was a perfect example of planting without the future in mind. If you're not going to stay on top of the pruning, space things farther apart.

It might make sense to add in some nitrogen-fixing and biomass creating trees at this point as well. I did not add any to my system because I had so many fruit I wanted to plant, but the extra soil improvement and biomass would be a good addition. Consider trees that can be aggressively pollarded for firewood, mulch and compost, like cassias, mimosa, *Enterolobium (spp.)*, moringa, Siberian pea shrub and others. I'm tempted to experiment with Paulownia and black locust trees as well, but they can propagate via roots and may become a nightmare. Instead of biomass trees we're growing lots of shorter cover crop plants and chopping and dropping prunings and spent plants on the beds right now, but I wouldn't mind having more organic matter. It's a balance, though, as many biomass creating plants are aggressive and also do not produce food. Further experimentation must be done. Currently, I think it makes sense to grow extra biomass outside the gardens to be brought in as mulch. If you like, you can just grow some short chop-and-drop plants like comfrey around the base of your trees and dedicate the more open space between the trees to the production of other crops. It's up to you. Bonus points if you get multiple uses out of your biomass-creating plants. I try to "stack functions", as we say in permaculture, getting as many possible uses out of a plant as possible. If a plant makes biomass, attracts insects, works well as a ground cover and is edible, it's got a lot of functions. If it's only a pretty ground cover, it's less useful and is better in a decorative planting than in a productive system like a Grocery Row Garden.

Once your trees are in, it's time to plant some shrubs and other perennials.

Step 4: Plant Your Perennials

In the gaps between the trees down the middle of the beds I planted a variety of small fruit and other perennial shrubs, including raspberries, mayhaws, thornless blackberries, rabbiteye blueberries and chaste trees. I put a single shrub in the gaps between fruit trees. You could go tighter, but again, you'll eat up the space you're going to need for growing vegetables. The idea is to ride the line, creating a forest edge where both annuals and perennials thrive, allowing you to harvest beans and apples, raspberries, asparagus, potatoes and peppers, etc., all from the same bed. Around and in between the shrubs, I added some patches of shorter-term perennials, like strawberries, cassava, yacon, chaya and asparagus. We also planted some canna roots around some of our fruit trees to use as a chop-and-drop plant. The species that has worked the best is *Canna musifolia*, which makes lots of gigantic, banana-like leaves. Though some gardeners have reported having chopped-and-dropped cannas root from the stems, this has never happened for me, despite weeks of wet weather which would have been ideal for rooting. Perhaps some other species do that and my *C. musifolias* don't. We also planted a few stems of *Tithonia diversifolia* here and there to use as chop-and-drop. In the future I'll also add moringa for the same reason, as well as for its nutrient-rich leaves. It's a good double-duty tree, since it's edible and it's a great chop-and-drop.

If you live farther north, patches of Jerusalem artichokes in the beds might work well. They produce a good bit of biomass you can use for mulch. Comfrey is also a great idea, as it's a very good chop-and-drop plant that will pull up nutrients and feed the soil.

In the dappled sunlight beneath the trees, consider planting species that like more shade. We've planted gingers in a couple of the beds as they appreciate a little break from the sun's heat. Farther north you might plant ramps, or even ginseng. If there's a plant that likes a little more shade, tuck it up under the trunk of a fruit tree.

Step 5: Plant Annuals

Once you've planted your perennial layers, it's time to plant the annuals. This mostly represents the herbaceous/groundcover layer in a food forest.

So far we've grown onions, collards, peppers, tomatoes, daikons, mustard, potatoes, tomatillos, sweet potatoes, pak choi, cabbages, beans and lots more. In Grenada we had piles and piles of sweet potatoes in our Grocery Row Gardens. Here I am planting them again, as they love to cover the ground and grow right through the heat of summer. (Though technically a perennial, in the US sweet potatoes are functionally an annual, like peppers, tomatoes and other short-lived perennial species that succumb to the cold of winter.)

In late winter I started multiple flats of vegetables in my little greenhouse which I subsequently planted out into the Grocery Row Gardens at regular spacing, fitting them in around the perennials. If there was a gap, I filled it with a transplant. We also direct seeded into some gaps with beans, summer squash, okra and Jamaican sorrel.

Once an annual is done, pull the plant and chop it into mulch for the garden, then plant something else. Walking around your Grocery Row Garden with a pocketful of seeds is a Good Gardener Best Practice (TM).

In the cool part of the year, go heavy on cool season vegetables, just like in a "regular" garden. In the spring after all danger of frost, switch to the heat-lovers. Just keep planting the gaps and you'll have food for a big chunk of the year.

In January we started the party off by planting white potatoes around all the beds, followed in February by broccoli, mustard, cabbage and collards. At the end of March we started planting tomatoes, followed by beans, peppers, eggplants and summer squash. In the tropics we just planted year round as gaps appeared. Beans would finish and we'd pop in tomatoes. Tomatoes would finish and we'd plant watermelons. It just kept on going and going. Do the same as your climate allows. If there's a gap, fill it. Right now I'm chopping pieces of sweet potato vines and planting them here and there any time a gap opens. Out with a rotting summer squash vine—in with some new sweet potato slips. I also recommend carrying beans around in your pockets. When you see openings, plant some beans in them. They'll feed the soil and give you food. You can also plant winter rye, buckwheat, clover and other good soil-builders in the gaps if you're not ready to plant any vegetables, then chop them down when you're ready to grow again.

Lay those gardens out, plant your trees, perennials and annuals, then keep the system going with new annuals as the old ones fail.

The soil improves over time if you keep dropping material onto it to rot and popping in plant life. The combination of humus and the root exudates from your living crops attract life. Worms, fungi, bacteria and other organisms thrive and minerals become more available. Dirt turns into soil.

Have fun with your plantings, above all. Don't get too tied into systems. If you like a plant and want to add it, add it! Throw in some caladiums for color if you like, or an *Amorphophallus* for sheer weirdness. Got a potted butterfly plant laying around? Plant it! Why not? Enjoy the beauty and pack it in. The more species you have, the more confused the pests will become. You'll also give predators a place to hide. I've seen a lot of baby toads, dragonflies, butterflies, bees, wasps and beetles making themselves at home in my Grocery Row Gardens and I welcome them.

Now that you've planted your perennials and your annuals, you are officially a Grocery Row Gardener. As you dream of future harvests a couple months down the road, let's move on and see why this method of gardening is such a winner.

Chapter 2

Why Grocery Row Gardens Work

Let's look at more of the benefits of this system.

Edge

I mentioned the power of a forest edge in the last chapter. This is an important permaculture concept. Edges are boundary lines between ecosystems. When you stand in a field, there is a decent amount of diversity. You'll find grasses, annuals, biennials and some perennials. Inside a forest there is also a decent amount of diversity. There are hickories and oaks, poplars and pines, doing most of their living at the canopy level. But at the overlap between ecosystems there is an explosion of life. You have both field and forest species. The trees aren't dominant and neither are the grasses and annual weeds. You get species from both ecosystems growing and interacting at the edge, not in too much sun or too much shade, not strangled by grass roots or overwhelmed by oaks. It's a transition zone between two different environments, with plants and animals from both. The shore is another classic example. There you'll find beach-dwelling species and aquatic species interacting. A massive variety of fish and animals and plants live in that little strip between the land and the sea. Crabs and clams, sea grasses and beach morning glory, mangroves and minnows, large fish and small fish, birds and bikini babes walking poodles. The beach is a happening place. Add a freshwater river entering the ocean and you get even more interactions between fresh and salt species. It's a free-for-all, a carnival of life. But go a little farther out or in and the number of species decreases.

A Grocery Row Garden is a set of edges. You have both forest and field species, with a mixture of full sun and part shade. You have open gaps to let in light and air and rain, but you also have denser areas. Some crops appreciate the light shade beneath a well-pruned tree. Others will reach for the edges of a bed. Grocery Row Gardens become strips of forest edge in a very short period of time. With your pruning and planting adding some disturbance and letting in light, they become even more productive.

Habitat

If you plant a range of species, you'll soon see bees and birds, wasps, spiders, praying mantises, ladybugs, beetles, dragonflies, butterflies and more. Life has found the rows. Provide a lovely space and they'll move in. As the system grows in diversity, pest problems decrease. Cabbage moths were a huge problem for us early in the year as the system started, but wasps moved in to eat them later on. Don't freak out about pests. Chances are, they'll pass in time. A robust ecosystem has a way of balancing itself. There are a lot of hiding spaces for the good guys in a Grocery Row Garden.

Increasing the habitat available is even better. I met an Alabama gardener named Ronnie Lehman who has at least a dozen birdhouses around his yard. Think of how many bugs those birds must eat, not to mention the high-quality manure they drop into his yard. It would make sense to put in birdbaths and birdhouses around your Grocery Row Garden. There will be produce to share, and they'll help feed the system for you by foraging for insects and then manuring your trees.

Wasp houses are also a good idea. Give them places to live and they'll happily destroy all the caterpillars they can catch. A bee hive is another good idea. Imagine all the good work they'll do pollinating your plants. You might also put small piles of rocks and sticks into your beds so good guys can take up residence. Or build an insect hotel. Or put in some tall poles for birds to sit on and watch your gardens for tasty pests. I hammered a few poles in the ground in Grenada and the kingbirds took that as an invitation to hang out and hunt bugs in the garden. We'd sit at our outdoor table for meals and watch them swoop off the poles after insects.

Make space for birds or other animals. At the very least, don't spray bug-killing poisons around. Say no to pesticides and you'll allow life to get a foothold and balance out the potential pest issues. I had a patch of heirloom corn in my garden that I didn't spray with anything. Despite it being a monoculture, I found ladybugs and praying mantises in there, along with beetles and plenty of spiders. Imagine how many more species will happily live in a Grocery Row Garden where there is a lot more diversity than in a corn patch! Keep away from the spray and life will come your way.

Polyculture

Habitat creation also ties into the concept of a polyculture. Monoculture plantings attract pests as they aren't a natural system. That isn't to say I don't enjoy doing some single-row gardening or planting patches of a single crop—I do—

but I also know that doing that again and again is a recipe for disaster. Florida's citrus industry is a prime example. Much has been made of the idea of companion planting, matching up various species in gardens that supposedly "like" each other. I think this is mostly ridiculous, but it's based on truth. I'm not going to go through a list to see what my celery "loves", but I have noticed that the more things I plant together, the fewer serious problems I have. Plants function best as part of a community of species. The pest that eats your tomato usually won't touch your dill, and the caterpillars on your dill would starve if you tried to feed them strawberries. Mixing up your plants mixes up the pests and heads off infestations. It also seems that different species of plants don't fight as much over space and seem to be happy planted in masses of growth, still happily yielding a crop. If you cram a bunch of the same species together, they fight over the same resources and don't give you as good a yield.

Trees Get Lots of Care

Let's face it: most of us don't take very good care of our trees. We might baby a bed of lettuce, giving it regular water and compost, but we don't do the same for a newly planted peach. Trees don't complain like annuals do. They're pretty tough and will often trudge along without much care, slowly growing bigger from year to year until they yield. This is less than ideal. When given plenty of water and good soil fertility, your trees grow like rockets, bearing better fruit in a much shorter period of time. Add trees to your garden beds and they get all the care that your vegetables get. If the melons are wilting, you'll see it and water them—and the apples get watered at the same time. If you don't feed the kale, it looks pathetic. When you see its yellow leaves and feed it, the tree gets some too. A Grocery Row Garden is a great place to be a tree. I have some trees growing by themselves in my front yard and they aren't nearly as happy as the ones I planted in my Grocery Row Gardens.

The System Is Convertible

Because of the original spacing of your fruit trees, if you ever get tired of growing vegetables and maintaining gardens, you can just let the trees become a regular orchard. Not that you'd ever want to, but there could be a time for that—like if you're selling your place to a boring non-gardener. Because we're currently renting our property, I thought of the possibility that one day the Grocery Row Gardens might end up as a regular orchard if untended. Once the trees are

Yields From The Beginning

Unlike an orchard, a Grocery Row Garden provides you with yields almost from the beginning. Though some of your pokier fruit trees, like pears, might take a half-decade to yield, your strawberries and beans don't. You'll be eating from your gardens in two months, and every year your yields will increase. In year one you'll get some berries along with lots of vegetables and roots. In year two you'll get vegetables and roots with a lot more berries and maybe a few peaches. In year three you'll have a variety of fruits and tons of berries with your roots and vegetables. As the years go on, you'll harvest more and more and more. You start with a yield and it just gets better, instead of starting with just non-productive trees and waiting for years.

Vertical Gardening Is Powerful

There have been a lot of articles on vertical gardening, mostly involving arbors and trellises and other infrastructure. Instead of dealing with all that, you can grow vertically in a Grocery Row Garden just by planting plants that grow at different heights. You get yields at head height, waist height, knee height, ground level and below the ground. It beats a bed of just beans or turnips hands down! You're getting that airspace to be productive instead of having everything down at the ground. You can also throw in trellises if you like (which I'll talk about later), but really, you're already getting some of the benefits of vertical gardening just based on the design of the system. As the trees grow they can also be used as supports for other species, such as beans, true yams, tomatoes and maypops.

No-Till

After the initial tilling and design of the system, a Grocery Row Garden is mostly no-till. It's hard to till around strawberries and plums and asparagus, so you can't really do it anymore. All you really have is some limited disturbance when you tear out some old plants and put in new ones, or if you rake an area up to plant a cover crop in a gap, or when you fill a trench full of squirrels that somehow died of lead poisoning in your heirloom corn patch. If you fear for the fungi and weep over the death of bacterial cultures every time you see a tractor turn the ground, your days of weeping are over. Grocery Row Gardens are places of love and kindness for all creatures. Except, occasionally, squirrels.

Regenerative Agriculture

The idea of regenerative agriculture ties in well with Grocery Row Gardening, as practiced and taught by Gabe Brown and others.

Brown's "Five Principles of Regenerative Agriculture" are:

1. Armor on the soil surface.
2. Least amount of chemical and physical disturbance possible.
3. Diversity of plants and animals, including insects.
4. Living roots in the soil as long as possible throughout the year.
5. Animals integrated into the system.

Read more at *greencover.com/introduction-to-regenerative-agriculture/*.

Since livestock would be hard to integrate into this system, we can make do on #5 by feeding weeds and vegetable waste to chickens and cycling their manure/compost back into the system. We can also encourage wildlife to be at home in our gardens, which integrates animal life into the ecosystem. As for #1, we have green cover and chop-and-drop mulch; for #2, we don't spray pesticides; for #3, we plant everything our hearts desire and more, bringing in lots of life; and for #4 we keep cover crops growing in between harvests. There is a good bedrock beneath this system, even if it isn't practiced on a farm-sized scale.

A Varied Diet

Another great benefit of the Grocery Row Gardening system is that you provide yourself and your family with a varied diet. In a few rows you could easily harvest a dozen or more different foods, especially as the system develops. This evening my children were daring each other to take bites of a brilliant red cayenne pepper they'd picked in the garden. They ate the entire thing, with multiple glasses of milk and a lot of laughter. But that pepper wasn't the only thing at the table. There was also a rich soup containing beef and a wide range of vegetables from the gardens, including okra, tomatoes, summer squash, fresh herbs, eggplant, peppers and whatever else Rachel found. Alongside that were two big trays of roasted homegrown potatoes. The Grocery Row Gardens are fantastic for providing a varied diet. Just plant whatever you like as you find a gap and you'll always have something good to eat during the growing season. If you also re-mineralize the soil like I have been doing with Solomon's Gold, you'll get better tasting produce with exceptional flavor.

Chapter 3

Maintaining a Grocery Row Garden

Once your Grocery Row Garden is planted and growing, there are some regular tasks to perform at least weekly. Most of these things can be done in a few minutes here and there as you have time. The system is remarkably resilient, though the more you work at it, the more productive it will become.

Here are some of the tasks that must be done to keep your Grocery Row Garden shipshape.

Pruning

When you first plant your fruit trees, I recommend pruning them. Stone fruit does well pruned to about knee height to encourage branching into a vase shape. However, you might consider cutting the trees at perhaps 3' instead in order to allow for more headroom for vegetable growth beneath the trees. We are currently experimenting with multiple prunings. Prune in the dormant season to get the initial shapes on the trees and to correct obvious problems, then prune again in the summer to control the height of trees. Pruning during dormancy increases the vigor of a tree and leads to outrageous growth, but pruning around the solstice at the height of the growing season will decrease tree vigor and help maintain short trees. Don't be afraid to be brutal. Buy Ann Ralph's book *Grow a Little Fruit Tree* and follow it. Being timid will lead to overgrown trees that outgrow their beds. You don't want that and it's easy to prevent. If all else fails, chainsaw them.

Re-planting

As I mentioned before, when you see a gap, plant in it. Pop in green onions from the store, plant the dry beans from a bag of 12-bean soup, put in a new strawberry plant, stick a potato in the ground. You know what to do! Keep those rows productive. The extra ground cover feeds the soil. In the summer I love to cut sweet potato vines and pop them all over the place, filling gaps and acting as a living mulch which later produces lots of potatoes. I also grow black-eyed peas in the gaps for extra nitrogen. If you have a cold climate and can plant some winter wheat or something like that to give you a final bit of cover into the cold, do it. It doesn't matter if you eat it. Just plant to create life and keep gaps from growing weeds. In a warmer climate, just keep plants growing in the gaps throughout the year. This ties into my next point of maintenance.

Mulching/Chop and Drop

Cover crops can be cut down and used for mulch. Some of them can also be eaten. Good candidates include bush beans, mustard, daikons, winter rye, wheat, oats, red clover and buckwheat. Think of plants that make biomass, add humus and maybe even fix nitrogen. You can also bring in mulch for your beds if you like. I do not recommend hay or straw due to the potential of herbicides killing your gardens, but grass clippings, fall leaves, pine needles, tree company wood chips and even shredded paper work well. I have been known to throw pieces of cardboard over weedy areas and then toss down mulch over the top. All my bills and IRS notices also end up as mulch in the Grocery Row Gardens. Some perennials, like comfrey and the cannas I mentioned above, are also great to chop and drop.

Don't let anything become too dominant in the system. If you have something growing too close to something else, chop it down with your machete and use it for mulch. All your tree prunings and spent plants should be thrown right back onto the ground to rot. We have mushrooms coming up here and there in the Grocery Row Gardens right now which it a great sign. That means they're eating up all the trimmings and bits of wood in the beds. Cornstalks are a good mulch, as is the Sorghum-Sudangrass hybrid we're growing right now. We have some patches of it outside the Grocery Row Gardens which we can cut and carry as mulch. In Syntropic Agriculture, there are entire beds dedicated to chop and drop crops grown as mulch for the rest of the system. There is probably a place for that in the Grocery Row Garden system as well, provided you have the space. As we experiment and tweak we'll discover more of what we need.

For now we are providing mulch via scavenging the yard and growing some in separate patches, as well as doing cover cropping inside the beds and regularly pruning and dropping spent vegetable matter back into the beds. There are some pieces of crummy ground a little farther from the house where we grow extra biomass. So far that is working well. Remember that if you live in the city you can also scavenge hedge prunings and grass clippings and fall leaves from your neighbors. Just don't do it if they have ChemLawn or throw around Weed and Feed.

Feeding

Unlike many permaculture practitioners and back-to-earth gardeners, I don't draw a hard line between "organic" and "chemical" fertilizers, especially after having my gardens destroyed by manure a decade ago, which is considered a great organic gardening amendment. I have never had, say, 10-10-10, kill a garden bed when applied in moderation. My main consideration now is the nutrient density of the food coming from my garden. After many discussions with Steve Solomon and much experimentation, I am quite comfortable adding a pinch of pure sodium molybdate for molybdenum and a (tiny) pinch of copper sulfate for copper and sulfur. I've got no problem foliar feeding a struggling tree with some DynaGro, either. My preferences run towards organic and I don't believe in burning out the soil by just tossing on bagged fertilizer year after year, but I've been known to give plants a hit when they need it. The plants don't care much where you got their food. But don't skimp on the trace elements, either. I encourage you to feed your garden with a wide range of materials and a good collection of micronutrients. The flavor difference is incredible. Just using compost from your own land or just using bagged 10-10-10 is not the best.

We've used all kinds of things to feed the gardens, as you've probably read in my book *Compost Everything*, or seen on my YouTube channel or read on my blog at TheSurvivalGardener.com. Dave's Fetid Swamp Water, seaweed, buried animals, leftovers—almost anything can feed your garden. The important thing is to keep it fed. If everything is turning yellow, don't hesitate to lime or to add some nitrogen. The regular chopping and dropping and mulching and cover cropping helps a lot, but you'll need more than that to grow great vegetables unless you have exceptional soil. Use what you have and keep those plants happy.

Watering

With a small Grocery Row Garden, you can easily drag a hose around to keep the system watered. If you don't get good rainfall and/or have a large garden, however, I recommend setting up some PVC stand pipes with sprinkler heads to overhead irrigate the garden. If you have denser, non-sandy soil, you might also consider drip irrigation. I am allergic to plumbing and have sandy soil so this isn't an option for us now. Thus far, we've been managing our large Grocery Row Gardens with just a hose. If we ever get the well working, I'll put up stand pipes. Fortunately, we live in a high rainfall area. Your mileage may vary.

Trellising

There are many ways to add trellises into a Grocery Row Garden. You could run a wire from one end to the other, like in a grape vineyard, training trees and vines to it. I don't do this because I don't like having a long barrier in the middle of a bed, and it's too much work. Instead, I stick pieces of cattle panel into the ground with a T-post to hold them up, then grow vining plants on those. Pick gaps between your trees and shrubs and you can pop them in, then take them out again at the end of the season if you wish. You can also stick poles into the ground for growing yams or pole beans. It's also possible to prune some trees higher so they can function as living trellises. We grew some peppers and tomatoes at the bases of fruit trees so we could tie them up later. If you want to get really creative, you can make arches over the path or even build grape pergolas right down a bed. It's up to you. I have some grape vines I'll be experimenting with over the next year. Putting grapes at perhaps 6' high on a single wire would allow you to garden normally beneath them and get a yield of both vegetables and grapes in the same space. Grape vines are normally planted at 20' spacing on single wires. It sounds like a long run, but grapes grow very fast. Alternate spacings may be worth trying. You might even try growing grapes up single poles with some trellis at the tops and seeing what happens. Have fun—be creative—grow grapes and whatever other vines you like inside the beds. This system is wide open for experimentation.

Maintaining the Paths

I leave the paths in my Grocery Row Gardens bare dirt. We mulch the beds, but the paths are dirt. It's very easy to maintain a dirt path with a wheel hoe. A regular hoe works fine as well, but it takes longer. You could mulch the pathways but it takes more effort and a lot more material.

Weeding

Since Grocery Row Gardens are a wild, rambling, crazy mixed polyculture, weeds in the beds need to be pulled by hand. It's hard to hoe when you have a big mix of plants growing together semi-randomly. Early in the system they are more of a problem since weeds like to show up in open gaps. As plants fill in the beds weeds have a difficult time gaining a foothold. Pull weeds as you see them so they don't go to seed. We throw them in the pathways so they don't re-root in the beds, then rake up the paths and throw the weeds to the chickens. You can also compost them or turn them into a "weed tea" by putting them into a barrel of water to rot down for a couple of weeks. When they're nice and rotted and stinky, water your beds with the liquid. If you're in a drier climate or rain isn't expected for a few days, you can also just pull weeds and throw them back into the Grocery Row Gardens to rot down as mulch. If you do this when it's rainy they often re-root—but in a dry week, they'll make decent mulch and the nutrients are returned to the beds.

Do a little work here and there and your beds will do well. If you do more work, planting regularly and getting the weeds out, fertilizing and watering, they'll do even better and will bear more.

Harvesting

As funny as it seems, some people overlook the important task of harvesting. They'll establish a big garden and plant everything, then fail to pick things on time, leading to overripe cucumbers the size of blimps and inedible yellow crooknecks. Certain plants have short harvest times and need daily picking, like okra. Others are very subject to insect damage, like tomatoes. We pick the latter when they aren't fully ripe and bring them in to sit on the counter and ripen, otherwise they almost invariably are ruined on the vine. If you pick regularly many plants produce for a longer period as well.

It's a joy to go for a shopping trip in your Grocery Row Gardens. Take a few baskets (thrift stores are a great place to get baskets) and a toddler and go picking. We're often surprised by what we find. Sometimes we get a handful of raspberries hiding under a tomato vine or a huge eggplant we missed. Our youngest daughter likes to pick all the berries she can find, both ripe and unripe, though she stubbornly refuses to touch the cherry tomatoes.

As you pick, you'll find the basics of many delicious meals and dishes. One of my sons recently dehydrated a large basket of little Everglades tomatoes in our dehydrator, resulting in explosively flavored little dried tomatoes. It was incredible! For breakfast we regularly eat sauteed eggplant and peppers, potatoes and broccoli and whatever else we have, along with bacon or sausage and eggs.

The eggs from our young chickens have just started coming in so now we have even more to look forward to. The chickens also eat well from our Grocery Row Gardens. Though they aren't free-range birds, we bring them wheel barrows of weeds and second-rate produce to munch on and know that they're getting the same high levels of nutrition we are, which will make their eggs much better for us than store-bought eggs.

Tools

Finally, though I may have mentioned most of the tools I use in the system in passing, here are my main go-to implements in the garden.

- A wheel hoe for maintaining pathways
- A machete for everything else

That's it.

No, just kidding.

We also use a normal long-handled combination shovel for planting trees, rebar and string for marking out beds, longer rebar for garden stakes, pieces of cattle panel with T-posts as trellises, bypass pruners for pruning trees, a wheelbarrow for pulled weeds and compost, produce, mulch and small children, a two-handled T-post hammer for hammering in rebar and T-posts, a hard-tined landscape rake for raking up weeds and making seedbeds, a pump sprayer for occasional foliar feeding, five-gallon buckets for hauling amendments, a pair of scissors for harvesting peppers and okra and an assortment of thrift store baskets for gathering produce.

But my machete is truly my main implement. It's used for digging transplant holes, chopping up biomass, harvesting, weeding and more. It's amazing how much you can do with a machete.

Grocery Row Gardens take a little thought to create and maintain, but once the produce starts rolling in and you go picking down lush rows as butterflies dance overhead, I think you'll be hooked. I sure am.

Chapter 4

The Influences

Now that you know how to build and maintain Grocery Row Gardens, let's cover the intellectual genesis of the system. Many people like to brag about their original systems or present old ideas as if they just discovered them. According to King Solomon, this is vanity—and who am I to argue with Solomon? There's nothing new under the sun. I've jokingly called Grocery Row Gardening my "1000% Original System" but it's anything but. Many of you will recognize portions of the Grocery Row Garden system in the work of other people, so let's give credit to its intellectual fathers and mothers.

Andrew Millison

There's a great video from 2018 on the Oregon State University Ecampus YouTube channel featuring permaculture teacher Andrew Millison showing off "The Ultimate Fruiting Permaculture Hedge". Andrew does good work and is a great teacher. The edible hedge concept ties into Grocery Row Gardening very closely, as it's basically a strip of forest edge. I planted an edible hedge after watching his video, just because it was so inspiring.

Ann Ralph

Ann Ralph is the author of *Grow a Little Fruit Tree*, a delightful little book that will open your eyes to the power of pruning and multi-planting trees to keep their sizes under control. I had previously seen some of these methods demonstrated by the guys at Dave Wilson Nursery and was not surprised to find that she had also worked with them. Her book took what I had observed in the past and solidified it into practical pruning advice that is easy enough for anyone to follow. It's one thing to hear from my friend Guy that he had luck "chainsawing his mulberry trees to the ground every three years," and another to find out exactly how that kind of pruning can work to maximum benefit in your backyard.

Bill Mollison

Bill Mollison was the co-founder of permaculture, which is a long-term system of gardening that is based on natural systems. He had a powerful combination of practicality and vision. When I read *Permaculture II*, I was struck by how incisive his observations were. They feel like they should be perfectly obvious, yet it took someone with great discernment to bring them to light. From Bill I began to learn about microclimates and edges, zones and design.

Craig Hepworth

Craig and I met in North Florida at Kanapaha Botanical Gardens almost a decade ago. His inquisitive nature and experimental bent inspired some of my food forest experiments. He also introduced me to the great value of *Dioscorea* yam species as a Florida survival crop, for which I remain indebted.

Dave Wilson Nurseries

About a decade ago the guys at Dave Wilson Nurseries put out a series of videos demonstrating various bizarre methods of keeping fruit trees small via a system called "Backyard Orchard Culture". I watched in fascination as they planted three or four fruit trees in a single hole and demonstrated just how short you could prune fruit and still get great yields. Instead of towering trees bearing fruit overhead, they had multiple apple varieties growing into a single bush-sized apple tree short enough for a child to manage. It is exciting stuff.

Eric Toensmeier

Eric Toensmeier's books *Perennial Vegetables* and *Paradise Lot* introduced me to a wide range of useful species which I had not previously encountered. I spent many hours reading and re-reading the plant profiles in the back of *Perennial Vegetables*, eventually acquiring and planting many of them in my own gardens. *Paradise Lot* was more the story of a pair of friends who buy a house together and transform a crummy urban lot into a permaculture paradise, finding joy—and wives—along the way. It's also inspiring.

Ernst Götsch

Ernst Götsch's Syntropic Farming was introduced to me late in my gardening journey but was a great encouragement as I developed my Grocery Row Gardening system. Because of him I added bananas to all my beds in Grenada, and learned how valuable the trunks can be for holding onto water, making paths, and for using as mulch inside beds. I am in awe at the intricacy of his system and how he uses succession to build the soil and create an ongoing yield. The tropics are already abundant, but he overclocked them and added rocket boosters! His long-term success assures me that what we are doing will work as well.

Geoff Lawton

Geoff Lawton's film *Establishing a Food Forest the Permaculture Way* first introduced me to how beautiful a food forest system could be. Though he was in tropical Australia and I was in Tennessee when I saw the film the first time, what I learned changed how I thought about trees and launched my decade of food forest planting.

Guido Marcelle

Dr. Guido Marcelle is the co-author of *Caribbean Spice Island Plants* and a personal friend of mine. He has built an impressive collection of tropical plants on his land and has a mind like a dictionary. He was of great help to me as I built my first Grocery Row Gardens in Grenada, identifying species I hadn't seen before and sharing a wide range of plant material, including a wild *Musa balbisiana* banana variety of great height, which I miss intensely.

John Jeavons

John Jeavons is the author of *How To Grow More Vegetables* and a leader in the Biointensive gardening movement. Though I rejected most of his approach to gardening and embraced those of his nemesis Steve Solomon, I will always be grateful for his work breaking me free of beds with boundaries. His gardening method of working right in the ground with light, fluffy well-dug beds has been a big help and a savings over the years.

Julia Morton

The late Julia Morton is my favorite gardening author. She introduced me to many wonderful plant species in her book *Fruits of Warm Climates*. I own most of her books in hardcover and highly recommend her writing. She was a brilliant researcher and a very clear writer who obviously loved the plants she wrote about.

Justin Rhodes

Justin Rhodes of *Permaculture Chickens* fame has been an internet friend of mine for years. After his YouTube channel took off, he encouraged me to build my own channel up, which I did, and he sent me quite a few new viewers. Due to this encouragement and the wider audience, I was then exposed to more good ideas from those that watch my videos and share their thoughts. Having the Youtube community accelerated my gardening experimentation as well as encouraged me to develop this system with more rapidity than

I would have otherwise managed. For that springboard I remain indebted to Justin.

Mark Shepard

Mark's book *Restoration Agriculture* presents the why and how of moving away from monoculture farms into resilient perennial systems. Instead of vast tilled fields growing only a single species, he reimagines agriculture as a living web of plants and animals, providing food over decades and centuries instead of a single season. I'm with him. Even though Grocery Row Gardens are basically a backyard gardening system and not something I developed for large farms, it's a great way to take a monoculture (a grass lawn!) and turn it into a long-term productive island of life. Restoring the land you own is a great step towards good stewardship of the earth we have been given.

Paul Wheaton

I count Paul Wheaton—the bad boy of permaculture—as a kindred spirit in many ways. We've been internet friends for quite a while, supporting each other's projects. The forums at permies.com have been an ongoing source of inspiration, especially years ago when I was drinking from the permaculture firehose.

Robert Hart

Englishman Robert Hart pioneered the modern forest gardening movement, building a temperate climate food forest for his disabled brother. He first observed and named the layers which we now take for granted. There is a strange video of Robert as an old man in his garden with his psychoanalyst. I have no idea why the psychoanalyst got into the action, but her being there has been a source of much amusement over the years.

Rosalind Creasy

Rosalind Creasy's books on edible landscaping encouraged us to see the beauty of edible plants, blurring the lines between decorative and utilitarian plantings. Though my Grocery Rows look little like Creasy's lovely edible landscape plots, I do pay much more attention to aesthetics than I did in the past, in part due to her influence.

Stefan Sobkowiak

Stefan Sobkowiak's film *The Permaculture Orchard* is a must-see. He takes a lot of theory and works it into a useful and profitable system that brings in an income. Though his climate is vastly different from mine, he really embraces the concept of integrating homes for wildlife and mixing up species instead of trying to fight nature to maintain a monoculture system. It's one thing to know this sort of thing in theory—it's another to see it in action, working over decades.

Steve Solomon

Steve Solomon is a close personal friend who I discovered long ago through his book *Gardening When It Counts*. We connected online and became friends, talking almost weekly about gardening, soil, faith and life. His contrarian approach to gardening pulls on a combination of ancient knowledge and modern science. He uses simple hand tools but irrigates with sophisticated Israeli sprinklers. He spaces his crops wide like a settler on the open plains, but buys individual elements through the mail for his intricate micronutrient fertilizers. He embraces the use of compost and biochar and eschews most pesticides while also using high-end hydroponic chemical fertilizers to create vegetables with maximum nutrition. I have been greatly influenced by his recommendations on growing nutrient-dense food as well as his ability to garden without irrigation. There's a reason I dedicated this book to him, even though my gardens look nothing like his.

Toby Hemenway

Toby Hemenway's book *Gaia's Garden* was a great influence on me when I read it years ago. Now I practice and preach many of the concepts in that book without even remembering where I originally found them. As I look back through its pages I remember the first time I read that book, page after page, excited with the possibilities. His book on home-scale permaculture is the best I have ever read on the topic, taking the high ideas of Bill Mollison and making them accessible to backyard gardeners.

I am certain there have been many more influences on this system that I missed, whether it was a link shared in an email, a lecture watched online, or a book I forgot to mention from my collection of gardening texts. If you gave me a suggestion along the road and saw me implement it, thank you. No man is an island and I owe a great debt to many other gardeners, authors and readers—especially to my favorite crew of Good Gardeners that are always sending me seeds and inspiration.

Chapter 5

Concluding Thoughts

As I mentioned at the beginning of this little book, the Grocery Row Garden system is young. For years I grew food forests and annual gardens separately and have only fully joined them in the last couple of years, so we don't have a long term track record on how this design will shape up in the future. I am sure we'll discover challenges and bonuses as it grows.

What we should see over the next few years is a ramp up in production from our perennials as the trees and shrubs mature. I also believe we'll see a further drop in pest activity as new life moves into the system. I would like to add bird houses on posts and some living area for wasps. Adding a birdbath is a good idea, and we might even put a beehive or two right into some of the rows. I would like to also play around more with trellises, maybe growing some single grape vines up individual posts inside some of the rows. Reducing the size of trees with larger growth potential so they fit inside the rows is another goal. We might try that with a pecan or two for an interesting challenge. Maybe if we keep chopping the main leader we can keep it under control. Who knows? It will also be very interesting to see how others work and expand upon this system. The framework is sound and there are thousands of plants that could be fit into it.

Tropical Grocery Row Gardens could be particularly productive, containing palms, ice cream bean, jabuticaba, plantains, mangos, starfruit, acerola cherries, soursop and more. With proper pruning, a massive collection of tropical fruit trees could be maintained inside this system. You might even be able to keep a jackfruit or a breadfruit controlled enough to work. It would be a marvelous experiment. Trees can be pruned and shaped much more than most people think possible. Just look at bonsai trees. You can make trees work for you in fascinating ways. I would also like to know how Grocery Row Gardens do in cold temperate climates. What species would you add? I could see apples and lingonberries, raspberries and cranberries, pie cherries and American plums, Jerusalem artichokes and fava beans. It would be amazing to see. You'd have more down time in the winter when snows came, but the garden would burst into life again in the spring,

CONCLUDING THOUGHTS

with that exuberant glory of spring growth that only happens after a cold winter.

As you join me in creating and testing your own Grocery Row Gardens, please let me know how it turns out. We can work on this system together, and when we've got all the kinks worked out, I can turn this little book into a nice, big book with plenty of illustrations. Thank you for reading, and until next time, may your thumbs always be green.

About the Author

David The Good is a gardening author and teacher who embraces the motto "more food for less work." He currently lives in Lower Alabama with his wife Rachel and their ten children, growing most of the family's produce and experimenting with a wide range of crops and techniques.

Find his writing on the web at thesurvivalgardener.com and see his gardening demonstrations at the popular "David The Good" YouTube channel.

www.ingramcontent.com/pod-product-compliance
Lightning Source LLC
Chambersburg PA
CBHW021430070526
44577CB00001B/152